THE ARITHMETICAL PRIMER.

UNDERHILL'S

NEW TABLE-BOOK;

OR,

TABLES OF ARITHMETIC MADE EASIER.

By D. C. UNDERHILL,
FORMERLY TEACHER IN FRIENDS' SCHOOL, NEW YORK.

A NEW EDITION,
REVISED, ENLARGED, AND IMPROVED.

NEW YORK:
CHARLES COLLINS, Publisher,
THE BAKER & TAYLOR CO., 740 BROADWAY.

PREFACE.

UNDERHILL'S NEW TABLE-BOOK, which, under one name or another, has for so many years enjoyed popular favor, is now presented to the public in an enlarged and improved form, and in a more worthy style of typography.

Alphabetical primers are no novelty; the present book, in its various editions, has been as yet the first and only attempt to produce an *arithmetical* primer fitted to initiate the young into the mysteries of figures, and to render the subject of numbers, in their uses and applications, attractive rather than repulsive to children.

A number of new pictorial illustrations will be found in the present edition. The Tables of Money, Weight, Measure, &c., have also been inserted in verse as well as in the standard form, in order to fix the facts permanently in the mind of the learner, so that they, either from memory or association, may be as promptly reproduced, when needed, as that universal one which recalls the number of days in each month.

It is only necessary to add that this little book is suited to the capacity of every child who is able to read, that it is designed to meet the wants of schools in every section of the country, and is equally well-adapted for home or parental instruction.

C. C. SAVAGE, STEREOTYPER,
13 *Chambers st., New York*

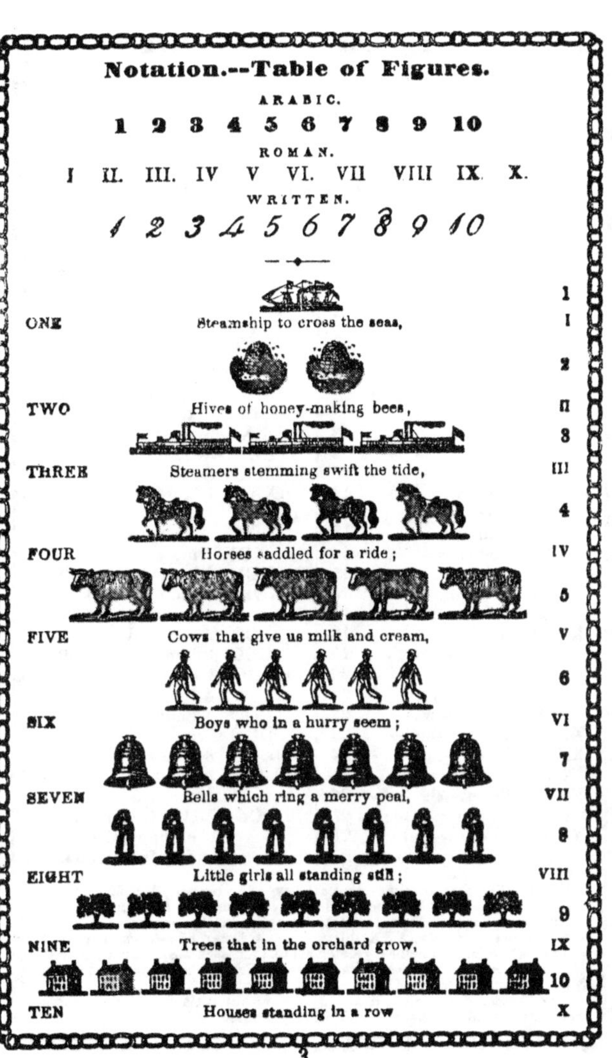

Notation.--Table of Figures.

ARABIC.

1 2 3 4 5 6 7 8 9 10

ROMAN.

I II. III. IV V VI. VII VIII IX. X.

WRITTEN.

1 2 3 4 5 6 7 8 9 10

		Arabic / Roman
ONE	Steamship to cross the seas,	1 I
TWO	Hives of honey-making bees,	2 II
THREE	Steamers stemming swift the tide,	3 III
FOUR	Horses saddled for a ride;	4 IV
FIVE	Cows that give us milk and cream,	5 V
SIX	Boys who in a hurry seem;	6 VI
SEVEN	Bells which ring a merry peal,	7 VII
EIGHT	Little girls all standing still;	8 VIII
NINE	Trees that in the orchard grow,	9 IX
TEN	Houses standing in a row	10 X

NUMERATION

OF

Roman and Arabic Figures.

Roman.		*Arabic.*	*Roman.*		*Arabic.*
I.	One	1	XXX.	Thirty	30
II.	Two	2	XXXI.	Thirty-one	31
III.	Three	3	XL.	Forty	40
IV.	Four	4	XLI.	Forty-one	41
V.	Five	5	L.	Fifty	50
VI.	Six	6	LI.	Fifty-one	51
VII.	Seven	7	LX.	Sixty	60
VIII.	Eight	8	LXI.	Sixty-one	61
IX.	Nine	9	LXX.	Seventy	70
X.	Ten	10	LXXI.	Seventy-one	71
XI.	Eleven	11	LXXX.	Eighty	80
XII.	Twelve	12	LXXXI.	Eighty-one	81
XIII.	Thirteen	13	XC.	Ninety	90
XIV.	Fourteen	14	XCI.	Ninety-one	91
XV.	Fifteen	15	XCV.	Ninety-five	95
XVI.	Sixteen	16	XCIX.	Ninety-nine	99
XVII.	Seventeen	17	C.	1 hundred	100
XVIII.	Eighteen	18	CC.	2 hundred	200
XIX.	Nineteen	19	CCC.	3 hundred	300
XX.	Twenty	20	CCCC.	4 hundred	400
XXI.	Twenty-one	21	D.	5 hundred	500
XXII.	Twenty-two	22	DC.	6 hundred	600
XXIII.	Twenty-three	23	DCC.	7 hundred	700
XXIV.	Twenty-four	24	DCCC.	8 hundred	800
XXV.	Twenty-five	25	DCCCC.	9 hundred	900
XXVI.	Twenty-six	26	M.	1 thousand	1000
XXVII.	Twenty-seven	27	M,DCCC,LIV.	One thousand eight hundred and fifty-four,	1854.
XXVIII.	Twenty-eight	28			
XXIX.	Twenty-nine	29			

NOTE.—The Roman characters are seven in number—I. V. X. L. C. D. M.—and any number may be formed by combinations of them.

The repetition of a letter repeats its value in the number. X being ten, XX make twenty, XXX thirty, &c. A letter of less value, placed at the right-hand of one of a greater value, increases the number; placed at the left-hand, it diminishes the number. Thus, V being five, and I one, VI is six, and IV four; XI eleven, and IX nine, &c,

The Roman mode of numbering is now but little used; the Arabic being preferred for business and other purposes.

Numeration.

1 2 3 4 5 6 7 8 9 10

How many balls are there in each line, counting three ways? Begin at 1, and count all around on the out side back to 1 again, and see if there are not twenty-seven Then count the whole, and see if there are not 55

Read the following numbers:—

45, 78, 67, 13, 46, 79, 35, 68. 14, 80, 36, 69, 15, 48, 81, 37.
70, 49, 82, 43, 76, 17, 50, 83. 38, 18, 51, 84, 39, 72, 19, 52.
85, 20, 53, 41, 74, 87, 42, 75. 22, 55, 88, 23, 56, 24, 57, 90.
11, 44, 25, 91, 77, 26, 59, 33. 92, 66, 27, 60, 93, 32, 65, 94.

NUMERATION TABLE.

The amount expressed by figures increases from right to left, but in reading or numerating them, commence at the left hand.

Hundreds of Sextillions Tens of Sextillions Sextillions	Hundreds of Quintillions Tens of Quintillions Quintillions	Hundreds of Quadrillions Tens of Quadrillions Quadrillions	Hundreds of Trillions Tens of Trillions Trillions	Hundreds of Billions Tens of Billions Billions	Hundreds of Millions Tens of Millions Millions	Hundreds of Thousands Tens of Thousands Thousands	Hundreds Tens Units
8 8 8,	7 7 7,	6 6 6,	5 5 5,	4 4 4,	3 3 3,	2 2 2,	1 1 1
Eight hundred and eighty-eight Sextillions.	Seven hundred and seventy-seven Quintillions,	Six hundred and sixty-six Quadrillions,	Five hundred and fifty-five Trillions,	Four hundred and forty-four Billions,	Three hundred and thirty-three Millions,	Two hundred and twenty-two Thousand,	One hundred and eleven.

NOTE.—The above is the French method, and the one generally used, the English method points off six figures to a period.

Table of Addition.

1	and	1	are	2	2	and	1	are	3
1	and	2	are	3	2	and	2	are	4
1	and	3	are	4	2	and	3	are	5
1	and	4	are	5	2	and	4	are	6
1	and	5	are	6	2	and	5	are	7
1	and	6	are	7	2	and	6	are	8
1	and	7	are	8	2	and	7	are	9
1	and	8	are	9	2	and	8	are	10
1	and	9	are	10	2	and	9	are	11
1	and	10	are	11	2	and	10	are	12
1	and	11	are	12	2	and	11	are	13
1	and	12	are	13	2	and	12	are	14
3	and	1	are	4	4	and	1	are	5
3	and	2	are	5	4	and	2	are	6
3	and	3	are	6	4	and	3	are	7
3	and	4	are	7	4	and	4	are	8
3	and	5	are	8	4	and	5	are	9
3	and	6	are	9	4	and	6	are	10
3	and	7	are	10	4	and	7	are	11
3	and	8	are	11	4	and	8	are	12
3	and	9	are	12	4	and	9	are	13
3	and	10	are	13	4	and	10	are	14
3	and	11	are	14	4	and	11	are	15
3	and	12	are	15	4	and	12	are	16
5	and	1	are	6	6	and	1	are	7
5	and	2	are	7	6	and	2	are	8
5	and	3	are	8	6	and	3	are	9
5	and	4	are	9	6	and	4	are	10
5	and	5	are	10	6	and	5	are	11
5	and	6	are	11	6	and	6	are	12
5	and	7	are	12	6	and	7	are	13
5	and	8	are	13	6	and	8	are	14
5	and	9	are	14	6	and	9	are	15
5	and	10	are	15	6	and	10	are	16
5	and	11	are	16	6	and	11	are	17
5	and	12	are	17	6	and	12	are	18

Table of Addition.—Continued.

7	and	1	are	8	8	and	1	are	9
7	and	2	are	9	8	and	2	are	10
7	and	3	are	10	8	and	3	are	11
7	and	4	are	11	8	and	4	are	12
7	and	5	are	12	8	and	5	are	13
7	and	6	are	13	8	and	6	are	14
7	and	7	are	14	8	and	7	are	15
7	and	8	are	15	8	and	8	are	16
7	and	9	are	16	8	and	9	are	17
7	and	10	are	17	8	and	10	are	18
7	and	11	are	18	8	and	11	are	19
7	and	12	are	19	8	and	12	are	20
9	and	1	are	10	10	and	1	are	11
9	and	2	are	11	10	and	2	are	12
9	and	3	are	12	10	and	3	are	13
9	and	4	are	13	10	and	4	are	14
9	and	5	are	14	10	and	5	are	15
9	and	6	are	15	10	and	6	are	16
9	and	7	are	16	10	and	7	are	17
9	and	8	are	17	10	and	8	are	18
9	and	9	are	18	10	and	9	are	19
9	and	10	are	19	10	and	10	are	20
9	and	11	are	20	10	and	11	are	21
9	and	12	are	21	10	and	12	are	22
11	and	1	are	12	12	and	1	are	13
11	and	2	are	13	12	and	2	are	14
11	and	3	are	14	12	and	3	are	15
11	and	4	are	15	12	and	4	are	16
11	and	5	are	16	12	and	5	are	17
11	and	6	are	17	12	and	6	are	18
11	and	7	are	18	12	and	7	are	19
11	and	8	are	19	12	and	8	are	20
11	and	9	are	20	12	and	9	are	21
11	and	10	are	21	12	and	10	are	22
[illegible]	and	11	are	22	12	and	11	are	23
[illegible]	and	12	are	23	12	and	12	are	24

Addition in Rhyme.

Two pennies had John,
His sister had 1,
They gave them to me,
And then I had 3,
Thus you may see,
That 2 and 1 make 3.

—

Two apples had Jane,
And Mary 2 more,
They gave them to Sarah,
And then she had 4;
Thus, 2 and 2 are 4, we know,
The apples make it plainly so.

—

James has 2 pears we see,
Then suppose I give him 3,
How many will there be?
2 and 3 are 5 we know,
So 3 and 2 for 5 must go,
Look in the table and find it so.

—

Margaret had a pincushion,
Presented by her mother,
It had 5 pins upon one side,
And 4 pins on the other—
On the little velvet ball,
How many pins were there in all?
5 and 4, as 9 we view,
And 4 and 5 are 9, as true,
The table will say the same to you.

Subtraction Table.

1 from 1 leaves 0	4 from 4 leaves 0
1 from 2 leaves 1	4 from 5 leaves 1
1 from 3 leaves 2	4 from 6 leaves 2
1 from 4 leaves 3	4 from 7 leaves 3
1 from 5 leaves 4	4 from 8 leaves 4
1 from 6 leaves 5	4 from 9 leaves 5
1 from 7 leaves 6	4 from 10 leaves 6
1 from 8 leaves 7	4 from 11 leaves 7
1 from 9 leaves 8	4 from 12 leaves 8
1 from 10 leaves 9	4 from 13 leaves 9
1 from 11 leaves 10	4 from 14 leaves 10
1 from 12 leaves 11	4 from 15 leaves 11
2 from 2 leaves 0	5 from 5 leaves 0
2 from 3 leaves 1	5 from 6 leaves 1
2 from 4 leaves 2	5 from 7 leaves 2
2 from 5 leaves 3	5 from 8 leaves 3
2 from 6 leaves 4	5 from 9 leaves 4
2 from 7 leaves 5	5 from 10 leaves 5
2 from 8 leaves 6	5 from 11 leaves 6
2 from 9 leaves 7	5 from 12 leaves 7
2 from 10 leaves 8	5 from 13 leaves 8
2 from 11 leaves 9	5 from 14 leaves 9
2 from 12 leaves 10	5 from 15 leaves 10
2 from 13 leaves 11	5 from 16 leaves 11
3 from 3 leaves 0	6 from 6 leaves 0
3 from 4 leaves 1	6 from 7 leaves 1
3 from 5 leaves 2	6 from 8 leaves 2
3 from 6 leaves 3	6 from 9 leaves 3
3 from 7 leaves 4	6 from 10 leaves 4
3 from 8 leaves 5	6 from 11 leaves 5
3 from 9 leaves 6	6 from 12 leaves 6
3 from 10 leaves 7	6 from 13 leaves 7
3 from 11 leaves 8	6 from 14 leaves 8
3 from 12 leaves 9	6 from 15 leaves 9
3 from 13 leaves 10	6 from 16 leaves 10
3 from 14 leaves 11	6 from 17 leaves 11

Subtraction Table.—Continued.

7 from 7	leaves	0		10 from 10	leaves	0
7 from 8	leaves	1		10 from 11	leaves	1
7 from 9	leaves	2		10 from 12	leaves	2
7 from 10	leaves	3		10 from 13	leaves	3
7 from 11	leaves	4		10 from 14	leaves	4
7 from 12	leaves	5		10 from 15	leaves	5
7 from 13	leaves	6		10 from 16	leaves	6
7 from 14	leaves	7		10 from 17	leaves	7
7 from 15	leaves	8		10 from 18	leaves	8
7 from 16	leaves	9		10 from 19	leaves	9
7 from 17	leaves	10		10 from 20	leaves	10
7 from 18	leaves	11		10 from 21	leaves	11
8 from 8	leaves	0		11 from 11	leaves	0
8 from 9	leaves	1		11 from 12	leaves	1
8 from 10	leaves	2		11 from 13	leaves	2
8 from 11	leaves	3		11 from 14	leaves	3
8 from 12	leaves	4		11 from 15	leaves	4
8 from 13	leaves	5		11 from 16	leaves	5
8 from 14	leaves	6		11 from 17	leaves	6
8 from 15	leaves	7		11 from 18	leaves	7
8 from 16	leaves	8		11 from 19	leaves	8
8 from 17	leaves	9		11 from 20	leaves	9
8 from 18	leaves	10		11 from 21	leaves	10
8 from 19	leaves	11		11 from 22	leaves	11
9 from 9	leaves	0		12 from 12	leaves	0
9 from 10	leaves	1		12 from 13	leaves	1
9 from 11	leaves	2		12 from 14	leaves	2
9 from 12	leaves	3		12 from 15	leaves	3
9 from 13	leaves	4		12 from 16	leaves	4
9 from 14	leaves	5		12 from 17	leaves	5
9 from 15	leaves	6		12 from 18	leaves	6
9 from 16	leaves	7		12 from 19	leaves	7
9 from 17	leaves	8		12 from 20	leaves	8
9 from 18	leaves	9		12 from 21	leaves	9
9 from 19	leaves	10		12 from 22	leaves	10
9 from 20	leaves	11		12 from 23	leaves	11

William had 3 peaches fair,
George and Thomas each his share,
One 8, the other 7;
Now search the table, and you will see,
That peaches 8 and peaches 3,
Make peaches just 11;
Add 7 peaches to 11, as the table shows,
And you'll have just 18, so the story goes.

Subtraction in Rhyme.

Three apples on a table lie,
And Jane takes one in passing by;
How many does she leave?
1 from 3 leaves 2,
So says the table true;
And 2 from 3 leaves 1,
When the work is rightly done;
And so must all believe.

Charlotte had 11 pins,
And gave Eliza 4;
Then all she had remaining
Were 7 and no more.
4 and 7 make 11,
Then 4 from 11 leaves 7,
And 7 from 11, four.

Peter has just 18 plums,
And gives his sister 10,
As gaily in the room he comes;
How many has he then?
The table makes it clear and straight,
That 10 from 18 leaves just 8;
And 8 from 18, ten—
From 18 plums take 10 away,
8 plums are left, 'tis plain as day.

Table of Multiplication.

2	times	1	are 2	5	times	1	are 5
2	times	2	are 4	5	times	2	are 10
2	times	3	are 6	5	times	3	are 15
2	times	4	are 8	5	times	4	are 20
2	times	5	are 10	5	times	5	are 25
2	times	6	are 12	5	times	6	are 30
2	times	7	are 14	5	times	7	are 35
2	times	8	are 16	5	times	8	are 40
2	times	9	are 18	5	times	9	are 45
2	times	10	are 20	5	times	10	are 50
2	times	11	are 22	5	times	11	are 55
2	times	12	are 24	5	times	12	are 60
3	times	1	are 3	6	times	1	are 6
3	times	2	are 6	6	times	2	are 12
3	times	3	are 9	6	times	3	are 18
3	times	4	are 12	6	times	4	are 24
3	times	5	are 15	6	times	5	are 30
3	times	6	are 18	6	times	6	are 36
3	times	7	are 21	6	times	7	are 42
3	times	8	are 24	6	times	8	are 48
3	times	9	are 27	6	times	9	are 54
3	times	10	are 30	6	times	10	are 60
3	times	11	are 33	6	times	11	are 66
3	times	12	are 36	6	times	12	are 72
4	times	1	are 4	7	times	1	are 7
4	times	2	are 8	7	times	2	are 14
4	times	3	are 12	7	times	3	are 21
4	times	4	are 16	7	times	4	are 28
4	times	5	are 20	7	times	5	are 35
4	times	6	are 24	7	times	6	are 42
4	times	7	are 28	7	times	7	are 49
4	times	8	are 32	7	times	8	are 56
4	times	9	are 36	7	times	9	are 63
4	times	10	are 40	7	times	10	are 70
4	times	11	are 44	7	times	11	are 77
4	times	12	are 48	7	times	12	are 84

Table of Multiplication.—Continued.

8	times	1	are	8	11	times	1	are	11
8	times	2	are	16	11	times	2	are	22
8	times	3	are	24	11	times	3	are	33
8	times	4	are	32	11	times	4	are	44
8	times	5	are	40	11	times	5	are	55
8	times	6	are	48	11	times	6	are	66
8	times	7	are	56	11	times	7	are	77
8	times	8	are	64	11	times	8	are	88
8	times	9	are	72	11	times	9	are	99
8	times	10	are	80	11	times	10	are	110
8	times	11	are	88	11	times	11	are	121
8	times	12	are	96	11	times	12	are	132
9	times	1	are	9	12	times	1	are	12
9	times	2	are	18	12	times	2	are	24
9	times	3	are	27	12	times	3	are	36
9	times	4	are	36	12	times	4	are	48
9	times	5	are	45	12	times	5	are	60
9	times	6	are	54	12	times	6	are	72
9	times	7	are	63	12	times	7	are	84
9	times	8	are	72	12	times	8	are	96
9	times	9	are	81	12	times	9	are	108
9	times	10	are	90	12	times	10	are	120
9	times	11	are	99	12	times	11	are	132
9	times	12	are	108	12	times	12	are	144
10	times	1	are	10	13	times	1	are	13
10	times	2	are	20	13	times	2	are	26
10	times	3	are	30	13	times	3	are	39
10	times	4	are	40	13	times	4	are	52
10	times	5	are	50	13	times	5	are	65
10	times	6	are	60	13	times	6	are	78
10	times	7	are	70	13	times	7	are	91
10	times	8	are	80	13	times	8	are	104
10	times	9	are	90	13	times	9	are	117
10	times	10	are	100	13	times	10	are	130
10	times	11	are	110	13	times	11	are	143
10	times	12	are	120	13	times	12	are	156

Multiplication in Rhyme.

Samuel has 2 knives,
 And Moses twice the same;
How many then for Moses,
 Ought we to name?
Twice means as many more,
Then Moses must have 4;
 Twice 1 are 2,
 Says the table to you:
 And twice 2 are 4,
 It says furthermore.

—

Amelia has 2 roses,
 And Frances has 2 more,
Miss Helen has another 2,
 To add unto the 4—
And 4 and 2 are 6 we call,
The number which they had in all:
So 3 times 2 are 6 we see,
And 6 we say for 2 times 3.

—

Four boys at marbles play,
And each has 5 they say;
How many marbles have they all?
Can any one the number call?
Yes, 5 and 5 we know are 10,
Two other fives the same again;
 And then 2 tens are 20;
So 4 times 5 do 20 make,
And 5 times 4 do 20 take,
And here are marbles plenty

Division in Rhyme.

Since 2 ones make 2 we know,
Then 2 but once in 2 will go;
Thus the father doth divide
2 apples, one to either side,
Which 2 good children share
As you can see them there.

Division Table.

1	in	2	goes	2	times	4	in	8	goes	2	times
1	in	3	goes	3	times	4	in	12	goes	3	times
1	in	4	goes	4	times	4	in	16	goes	4	times
1	in	5	goes	5	times	4	in	20	goes	5	times
1	in	6	goes	6	times	4	in	24	goes	6	times
1	in	7	goes	7	times	4	in	28	goes	7	times
1	in	8	goes	8	times	4	in	32	goes	8	times
1	in	9	goes	9	times	4	in	36	goes	9	times
1	in	10	goes	10	times	4	in	40	goes	10	times
1	in	11	goes	11	times	4	in	44	goes	11	times
1	in	12	goes	12	times	4	in	48	goes	12	times
1	in	13	goes	13	times	4	in	52	goes	13	times
2	in	4	goes	2	times	5	in	10	goes	2	times
2	in	6	goes	3	times	5	in	15	goes	3	times
2	in	8	goes	4	times	5	in	20	goes	4	times
2	in	10	goes	5	times	5	in	25	goes	5	times
2	in	12	goes	6	times	5	in	30	goes	6	times
2	in	14	goes	7	times	5	in	35	goes	7	times
2	in	16	goes	8	times	5	in	40	goes	8	times
2	in	18	goes	9	times	5	in	45	goes	9	times
2	in	20	goes	10	times	5	in	50	goes	10	times
2	in	22	goes	11	times	5	in	55	goes	11	times
2	in	24	goes	12	times	5	in	60	goes	12	times
2	in	26	goes	13	times	5	in	65	goes	13	times
3	in	6	goes	2	times	6	in	12	goes	2	times
3	in	9	goes	3	times	6	in	18	goes	3	times
3	in	12	goes	4	times	6	in	24	goes	4	times
3	in	15	goes	5	times	6	in	30	goes	5	times
3	in	18	goes	6	times	6	in	36	goes	6	times
3	in	21	goes	7	times	6	in	42	goes	7	times
3	in	24	goes	8	times	6	in	48	goes	8	times
3	in	27	goes	9	times	6	in	54	goes	9	times
3	in	30	goes	10	times	6	in	60	goes	10	times
3	in	33	goes	11	times	6	in	66	goes	11	times
3	in	36	goes	12	times	6	in	72	goes	12	times
3	in	39	goes	13	times	6	in	78	goes	13	times

Division Table.—Continued.

7 in 14 goes 2 times	10 in 20 goes 2 times
7 in 21 goes 3 times	10 in 30 goes 3 times
7 in 28 goes 4 times	10 in 40 goes 4 times
7 in 35 goes 5 times	10 in 50 goes 5 times
7 in 42 goes 6 times	10 in 60 goes 6 times
7 in 49 goes 7 times	10 in 70 goes 7 times
7 in 56 goes 8 times	10 in 80 goes 8 times
7 in 63 goes 9 times	10 in 90 goes 9 times
7 in 70 goes 10 times	10 in 100 goes 10 times
7 in 77 goes 11 times	10 in 110 goes 11 times
7 in 84 goes 12 times	10 in 120 goes 12 times
7 in 91 goes 13 times	10 in 130 goes 13 times
8 in 16 goes 2 times	11 in 22 goes 2 times
8 in 24 goes 3 times	11 in 33 goes 3 times
8 in 32 goes 4 times	11 in 44 goes 4 times
8 in 40 goes 5 times	11 in 55 goes 5 times
8 in 48 goes 6 times	11 in 66 goes 6 times
8 in 56 goes 7 times	11 in 77 goes 7 times
8 in 64 goes 8 times	11 in 88 goes 8 times
8 in 72 goes 9 times	11 in 99 goes 9 times
8 in 80 goes 10 times	11 in 110 goes 10 times
8 in 88 goes 11 times	11 in 121 goes 11 times
8 in 96 goes 12 times	11 in 132 goes 12 times
8 in 104 goes 13 times	11 in 143 goes 13 times
9 in 18 goes 2 times	12 in 24 goes 2 times
9 in 27 goes 3 times	12 in 36 goes 3 times
9 in 36 goes 4 times	12 in 48 goes 4 times
9 in 45 goes 5 times	12 in 60 goes 5 times
9 in 54 goes 6 times	12 in 72 goes 6 times
9 in 63 goes 7 times	12 in 84 goes 7 times
9 in 72 goes 8 times	12 in 96 goes 8 times
9 in 81 goes 9 times	12 in 108 goes 9 times
9 in 90 goes 10 times	12 in 120 goes 10 times
9 in 99 goes 11 times	12 in 132 goes 11 times
9 ln 108 goes 12 times	12 in 144 goes 12 times
9 in 117 goes 13 times	12 in 156 goes 13 times

Division in Rhyme.

Harriet has 4 oranges,
For little Jane and Mary,
In equal share, each has a pair,
The numbers do not vary.
And thus we see Division true;
Both have the 4, and each has 2;
2 twos are then in four we know,
And 2 in 4 will 2 times go.

—

Six large apples Henry had,
To give to Silas, John, and Thomas,
And 2 he gave each little lad,
According to his promise;
And as he dealt them round,
3 twos in 6 were found;
So 2 in 6 will 3 times go,
Then 3 in 6 goes twice we know.

—

A teacher had just 20 toys,
To give to 4 good little boys,
He would the fives in 20 teach,
And so he gave 5 toys to each,
4 fives in 20 thus he shows,
And then 5 fours, as each boy knows;
So 5 into 20 will go 4 times,
And 4 into 20 go 5, correctly chimes;
The table will show you the truth of the rhymes.

The Uses of the Tables.

Notation writes the figures down,
And *Numeration* reads them—
Addition makes two numbers one,
And more so, when it needs them.
Subtraction of two numbers makes
A third—as you have seen them—
The smaller from the larger takes,
And shows the odds between them,
Multiplication, in a word,
Adds much with little labor,
And with two numbers makes a third,
A far superior neighbor.
Division with two numbers shows
How many times attaining,
The less one in the larger goes,
And what there is remaining.
The large square table at a view,
Will show what each and all can do

TABLES OF

MONEY, WEIGHTS, MEASURES, ETC.

I. MONEY.

Money has various names, or rates:
FEDERAL, or that of the United States,
Has *Eagles*, *Dollars*, *Dimes*, and *Cents*.
STERLING, or English, which one sees
In England and her colonies,
Has *Guineas*, *Pounds*, *Shillings*, and *Pence*.
FRENCH, used in France through all her ranks,
Has simple *Centimes*, *Décimes*, *Francs*

100 Cents.

3 Cents

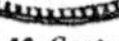
50 Cents

10 Cents

250 Cents.

500 Cents.

25 Cents.

FEDERAL, OR UNITED STATES MONEY.—The value of this money increases by tens, which makes it very simple and easy to reckon, and hence accounts are mostly kept in Dollars and Cents, in this country.

TABLE.

10 mills *m.* make	1 cent, *c.*
10 cents	1 dime, *d.*
10 dimes	1 dollar, $.
10 dollars	1 eagle, *E.*

Observe the denominations—Eagle, *E.*; dollar, $; dime, *d.*; cent, *c.*; mill, *m.*

The mill is not a coin, and is only used in counting.

Beside money made of silver, gold, and copper, there are also pieces of ornamented paper representing money, upon which are printed sums from one dollar to one thousand; these pass readily for what is stamped upon them for all the purposes of business.

We count 10 *mills* to every *cent*,
10 cents to every *dime;*
10 dimes are in the *dollar* spent,
10 dollars in the *eagle* chime,

6 Pence.

1 Shilling.

1 Sovereign.

ENGLISH OR STERLING MONEY.—This currency is used in England and the British colonies.

TABLE.

4 farthings, *qr.*, make	1 penny, *d.*
12 pence	1 shilling, *s.*
20 shillings	1 pound, £.
5 shillings	1 crown, *c.*
20 shillings	1 sovereign, *sov*
21 shillings	1 guinea, *G.*

Observe the denominations—Pound, £; shilling, *s.* pence, *d.*; farthing, *qr.*; from the Latin words, *libra*, a pound; *solidus*, a shilling; *denarius*, a penny; *quadrans*, a farthing.

Farthings are marked in fractions thus:—¼, one farthing; ½, two farthings, or half-penny; ¾, three farthings.

The sovereign, which is made of gold, is of the same value as the pound. Whatever costs a pound, therefore, a sovereign will pay for. There is no coin called a pound.

4 *farthings* make one English *penny;*
12 pence are in one *shilling* found:
While 21 shillings make one *guinea*,
And 20 shillings form one *pound.*

---

5 shillings make a silver *crown*,
A sovereign is a golden *pound.*

50 Centimes.

1 Franc.

2 Francs.

FRENCH MONEY.—This money is used in France, and also in the French colonial possessions to some extent.

TABLE.

10 centimes,* *c.*, make - -	1 décime, *d.*
10 décimes,† - - - -	1 franc,‡ *fr.*

Observe the denominations—Franc, *fr.*; décime, *d.*: centime, *c.*

The five-franc piece is frequently met with in this country, and passes currently at 94 cents. Coins of a less number of francs, pass also at the rate of 19 cents to the franc.

10 *centimes* in one *décime* meet,
10 decimes make one *franc* complete.

FEDERAL and FRENCH are decimal,
They count by ten alone:
Of coins it always take 10 small
To make the next larger one.
The ENGLISH currency, 'tis thought,
To the same standard will be brought.
The *guinea* then no more we'll see,
The *pound* 1000 mills will be;
And all the lower coins will range,
In just accordance with this change.

II. WEIGHTS.

TROY WEIGHT of silver and gold disposes·
APOTHECARIES' WEIGHT is for doctors' doses;
Whatever else your trade employs,
Comes under the rule of AVOIRDUPOIS.

---

* Pronounced *sonteen*. † Pronounced *daseem*.
‡ Pronounced *fraung*, or frank.

TROY WEIGHT—Is used to weigh Gold, Silver, Jewels, and Liquors.

TABLE.

24 grains, *gr*., make	1 pennyweight, *dwt.*
20 pennyweights - - -	1 ounce, *oz.*
12 ounces - - - -	1 pound, *lb.*

Observe the denominations—Pound, *lb.*; ounce, *oz.*: pennyweight, *dwt.*, grain, *gr.*

This rule for weighing gold, will state
That 24 *grains* make one *pennyweight:*
And 20 pennyweights in an *ounce* are found,
While 12 good ounces make a *pound.*

APOTHECARIES' WEIGHT.—Is used in mixing doses of medicine, but drugs and medicines, like most other merchandise, are bought and sold by AVOIRDUPOIS WEIGHT.

TABLE.

20 grains, *gr*., make	1 scruple, ℈.
3 scruples - -	1 drachm, ʒ.
8 drachms - -	1 ounce, ℥
12 ounces - -	1 pound, ℔.

Observe the denominations—Pound, ℔.; ounce, ℥.; drachm, ʒ.; scruple, ℈., grain, *gr.*

In mixing doses, Doctors say
That 20 *grains* one *scruple* weigh;
3 scruples make one *drachm* they held,
8 drachms are to the ounce enrolled,
And ounces 12 for a *pound* are sold.

AVOIRDUPOIS WEIGHT—Is used in weighing all coarse and heavy goods, groceries, &c.; and all metals except silver and gold.

TABLE.

16 drachms, *dr*., make -	1 ounce, *oz.*
16 ounces - -	1 pound, *lb.*
25 pounds - - -	1 quarter, *qr.*
4 quarters - - -	1 hundred weight, *cwt.*
20 hundred weight - -	1 tun, *T*

Observe the denominations—Tun, *T.*; hundred weight, *cwt.*; quarter, *qr.*; pound, *lb.*; ounce, *oz.*; drachm, *dr.*

2000 pounds make a tun in the table; when, as it sometimes is, it is 2240 lbs., 28 pounds make a quarter. Except in special cases, the tun is now regarded as 2000 pounds by the principal merchants of our cities.

16 *drachms* make one *ounce*,
16 ounces make one *pound*,
25 pounds one *quarter* counts,
4 quarters make a *hundred* round;
And 20 hundred weight are run,
To make a full and perfect *tun*.

'Twas once absurdly held and said
112 pounds make a hundred weight;
And then this table always read,
That a quarter hundred was 28

III. MEASURES.

By different measures, we obtain
Due quantities of wood or grain,
Of cloth, or land, or wine, and tell
How much of each we buy or sell.
Cloth Measure is for ribbons, tapes,
And cloths, and silk, for coats or capes.
Long Measure serves to tell and trace
The distances from place to place.
Surveyors' Measures, understand,
Are only used in measuring land.
Dry Measure tells how much we gain
Of salt, coal, fruit, potatoes, grain;
While Liquid Measure justly classes
Wine, spirits, beer, oil, milk, molasses.
Square Measure deals with surfaces,
As walls, and floors, and fields, and seas;
And Cubic Measure ascertains
What any solid shape contains.
Time Measure tells us, as they fly,
How days, months, years, are rushing by;
And Circular Measure shows the worth
Of lines that circle round the earth,
And of the bands which reason's eye
Traces across the glittering sky.

1. MEASURES OF LENGTH.

CLOTH MEASURE—Is used to measure Cloths, Ribbons, Tapes, &c.

TABLE.

2¼	inches, *in.*, make - -	1 nail, *n.*
4	nails - - - - -	1 quarter of a yard, *qr.*
4	quarters - -	1 yard, *yd.*

FOREIGN CLOTH MEASURES.

2½	quarters, make - -	1 ell Hamburgh, *E. H.*
3	quarters - - -	1 ell Flemish, *E. F.*
5	quarters - -	1 ell English, *E. E.*
6	quarters - - -	1 ell French, *E. Fr.*

Observe the regular denominations— Yard, *yd.*; quarter, *qr.*; nail, *n.*; inch, *in.*

In measuring cloth for use or sale,
2¼ *inches* make one *nail;*
4 nails one *quarter* we regard,
And four full quarters make one *yard.*
Nails now are seldom used, we've heard,
Eighths and sixteenths are much preferred.

LONG MEASURE—Is used to measure distances, and to ascertain the length of anything without regard to breadth.

TABLE.

10	lines, *l.*, make -	1 inch, *in.*
12	inches - - - -	1 foot, *ft.*
3	feet - - -	1 yard, *yd.*
5½	yards - - - -	1 rod or pole, *p.*
40	poles, or 220 yards - -	1 furlong, *fur.*
8	furlongs - - - - - -	1 mile, *M.*
3	miles - - - - - -	1 league, *L.*
60	geographic, or 69½ statute miles	1 degree, *Deg.*
360	degrees -	the circumference of the earth.

Observe the denominations—Degree, *Deg.*; league, *L.*; mile, *M.*, furlong, *fur.*; rod, or pole, *p.*; yard, *yd.*, foot, *ft.*; inch, *in.*; line, *l.*

Twelve lines make an inch in France.

In measuring the height of horses the *hand*, 4 inches, is used; and in measuring the depth of water, the *fathom*, 6 feet, is used.

In measuring distances or lengths,
10 *lines* are said to make one *inch;*
12 inches make a perfect *foot,*
3 feet into a *yard* are put;
5½ yards make a *rod* or *pole,*
And 40 rods a *furlong* whole;
8 furlongs make a *mile* quite big,
And 3 full miles make up a *league.*
In measuring round the earth, we see,
That 60 miles make one *degree;*
Degrees 360, then,
The *earth's circumference* will span.

SURVEYOR'S MEASURE.—This measure is used in ascertaining the length and breadth of land, roads, &c.

TABLE.

7 92-100 inches, *in.*, make - -	1 link, *l.*
25 links - - - - - -	1 pole, *p.*
4 poles, or 100 links - -	1 chain, *c.*
10 chains - - -	1 furlong, *fur*
8 furlongs - - -	1 mile, *M.*

Observe the denominations—Link, *l.*; pole, *p.*; chain, *c.*; furlong, *fur.*; mile, *M.*

7 *inches and ninety-two hundredths*, make
One *link* in the chain surveyors take,
100 links his *chain* embraces,
With 80 chains one *mile* he traces.

---

2. MEASURES OF CAPACITY.

LIQUID MEASURE—Is used in measuring Wine, Spirits, Beer, Oil, Vinegar, Milk, Molasses, &c.

TABLE.

4 gills, *g.*, make - - -	1 pint, *pt.*
2 pints - - - - -	1 quart, *qt.*
4 quarts - - - - -	1 gallon, *gal.*
31½ gallons - - - -	1 barrel, *bbl.*
2 barrels - - -	1 hogshead, *hhd.*
2 hogsheads - -	1 pipe, *p.*
2 pipes - - -	1 tun, *t.*

Observe the regular denominations—Tun, *t.*; pipe, *p.*; hogshead, *hhd.*; barrel, *bbl.*; gallon, *gal.*; quart, *qt.*; pint, *pt.*; gill, *g.*

MEASURES OF CAPACITY SELDOM USED.

TABLE.

9 gallons, *g.*, make - -	1 firkin, *f.*
10 gallons - - - - -	1 anker, *a.*
2 firkins, or 18 gallons -	1 kilderkin, *k.*
2 kilderkins, or 36 gallons	1 barrel of beer, &c. *bbl.*
1½ barrels, or 54 gallons -	1 hogshead of beer, *hhd.*
42 gallons - - - - -	1 tierce, *t.*
2 tierces, or 84 gallons -	1 puncheon, *p.*

Most liquids are now sold by the gallon, quart, and pint, and not by the other denominations of liquid measure; in fact, vessels are rarely made to hold the exact quantities stated in the table, and frequently retain the names though containing much more, the hogshead and barrel for instance.

In measuring liquids, first we take
4 little *gills* one *pint* to make;
2 pints then make one *quart*, and still
4 quarts the *gallon* measure fill.
Gallons one half and 31,
Will fill a *barrel* to the bung.
2 barrels to the *hogshead* go,
2 hogsheads fill a *pipe*, and so
2 pipes will near a *tun* o'erflow.

Though many *good things* are measured still
By gallon, quart, and pint, and gill,
Yet Liquid Measure oft seems to me
" The measure of human misery."
For O, what countless evils come
From brandy, whiskey, gin, and rum,
Which it were better ne'er to touch,
For a *single drop* is " a drop too much."

DRY MEASURE—Is used in measuring Grain, Potatoes, Fruit, Coal, Salt, Seeds, &c.

TABLE.

2 pints, *pt.*, make -	1 quart, *qt.*
8 quarts - - -	1 peck, *p.*
4 pecks -	1 bushel, *bush.*
36 bushels - - -	1 chaldron of coal, *chal.*
8 bushels - - -	1 quarter of corn, *qr.*

Observe the regular denominations—Chaldron, *chal.*; bushel, *bush.*; peck, *p.*; quart, *qt.*; pint, *pt.*

2 *pints* DRY MEASURE make one *quart*,
8 quarts one honest *peck* contains,
4 pecks are in a *bushel* brought;
8 bushels, if you are measuring grains,
Are to the *quarter** counted out;
But if bituminous coal,† you take
Then 6 and 30 bushels make
The *chaldron*, which in trade obtains.

---

3. MEASURES OF CONTENT.

LAND OR SQUARE MEASURE—Is used in reckoning the contents of any surface by its length and breadth.

TABLE.

144	square inches, *s. in.*, make	1 square foot, *S. F.*
9	square feet - -	1 square yard, *S. Y.*
30½	square yards - -	1 square rod, pole, or perch, *S. P.*
40	square poles - - -	1 square rood, *S. R.*
4	square roods - -	1 square acre, *S. A.*
640	square acres - -	1 square mile, *S. M.*

Observe the denominations—Square mile, *S. M.*; square acres, *S. A.*, square rood, *S. R.*; square perch, *S. P.*; square yard, *S. Y.*; square foot, *S. F.*; square inch, *S. I.*

Square inches one hundred and forty-four
Make one *square foot*, and nothing more.
9 square feet make one *square yard*,
30 yards and a quarter are one *pole squared*;
40 square poles make one *square rood*,
Yet 4 square roods make an *acre* good;
And acres 640 the while,
Are wanted to make up one *square mile*.

SOLID OR CUBIC MEASURE—Is used to reckon the contents of anything by its length, breadth, and depth.

TABLE.

1728 solid inches, *s. in.*,	make	1 solid foot, *S F*
40 feet of round, or 50 feet of hewn	timber	1 tun, *Tun.*
27 solid feet	- -	1 solid yard, *S. Yd.*
16 solid feet of wood	-	1 cord foot of wood, *Ft. W*
8 cord feet of wood	-	1 cord, *C.*

Observe the denominations—Cord, *C.*; cord foot of wood, *Ft. W.*, solid yard, *S. Y.*; solid foot, *S. F.*; solid inch, *S. I.*

---

* Wheat is measured in Great Britain by the quarter of 480 lbs.
† Bituminous, or soft coal, is sold by measure, and anthracite, or hard coal, by weight.

One thousand seven hundred and twenty-eight
Inches one solid *foot* complete.
In *timber*, 40 feet, if *round*,
Or 50 *hewn*, a *tun* is found.
In measuring ships the rule must run,
Feet 2 and 40 make a tun.
Feet 27 one *solid yard* we rate,
A *cord of wood* one hundred twenty-eight.

4. MEASURES OF DURATION AND CIRCULAR DISTANCES.

TIME MEASURE—Is used in computing the different periods in which any transaction or event occurs.

TABLE.

60 seconds, *sec.*, make	1 minute, *m.*
60 minutes	1 hour, *h.*
24 hours	1 day, *d.*
7 days	1 week, *w.*
4 weeks	1 month, *mo.*
12 months	1 year, *yr.*
100 years	1 century, *C.*

Observe the denominations — Century, *C.*; year, *yr.*; month, *mo.* week, *w.*; day, *d.*; hour, *h.*; minute, *m.*; second, *sec.*

12 calendar months, or 13 lunar months, 1 day, and 6 hours, or 365 days, 6 hours, 1 common, or Julian year.

The year is divided by the calendar as follows:—

			DAYS.				DAYS.
1st	month,	January, has	31	7th	month,	July, has	31
2d	"	February,	28	8th	"	August,	31
3d	"	March,	31	9th	"	September,	30
4th	"	April,	30	10th	"	October,	31
5th	"	May,	31	11th	"	November,	30
6th	"	June,	30	12th	"	December,	31
			181				365

February has 29 days every fourth year, which is called Bissextile, or Leap Year. Every Leap Year may be divided by 4 without a remainder; other years, divided by 4, leave one, which shows their number after Leap Year. Thus, 1854 divided by 4, leaves a remainder of 2: that is, it is the second after Leap Year, &c., &c.

The number of days in each month, may easily be remembered by the following verse:—

Thirty days have September,
April, June, and November;
All the rest have thirty-one,
Excepting February alone,
Which hath twenty-eight, nay more,
Hath twenty-nine one year in four.

There are in every year four seasons, viz: SPRING, SUMMER, AUTUMN, and WINTER.

The Spring months are March, April, and May.

The Summer months are June, July, and August.

The Autumn months are September, October, and November.

The Winter months are December, January, and February.

The Spring is the season of flowers; the Summer of fruits; the Autumn of the decay of vegetation and the fall of the leaf; and the Winter of frost and snow.

60 *seconds* make 1 *minute*,
  Time enough some good to do;
60 minutes make 1 *hour*,
  Who will dare to waste it? Who?
24 hours make up the *day*,
Time for work, and sleep, and play;
7 days to the *week* are given,
Six for toil and one for heaven.
God gives me six for work and play,
I will not steal the seventh away.
4 weeks in every *month* appear,
12 months make up the rolling *year*;
100 years—few live them to see—
Are what are called a *century*.

CIRCULAR MEASURE—is used by Astronomers, Navigators, &c in making their calculations.

TABLE.

60 seconds, ″ make - -	1 minute, ′
60 minutes - - - - -	1 degree, °
30 degrees - - - -	1 sign, *S.*
12 signs, or 360 degrees -	1 circle of the Zodiac, *C.*

Observe the denominations—Circle, *C.*; sign, *S.*; degree, °; minute, ′; second, ″.

60 *seconds* make one *minute*,
60 minutes one *degree:*
30 degrees one *sign* has in it,
12 signs we in { a circle / the *zodiac* } see.

'Tis knowledge gained from this, that guides
The ship, that o'er the ocean rides,
And shows the pilot how to steer
From place to place, remote or near.

IV. BOOKS, PAPER, AND PARCHMENT.

BOOKS, PAPER, PARCHMENT, all concern
Men of a literary turn:
As authors, printers, and booksellers—
A race of genuine clever fellows—
The paper manufacturer, too,
With these, of course, has much to do;
The stationer and binder then—
Known as industrious, thrifty men—
Bear each an honorable part
In the noble, intellectual art,
Of furnishing the mind and heart.

1. PAPER AND PARCHMENT

TABLE OF PAPER AND PARCHMENT.—This table is used by Papermakers, Printers, and dealers in Stationery, &c., &c.

TABLE.

24 sheets of paper, make	- -	1 quire, *qr.*
20 quires	- - - - -	1 ream, *Rm.*
2 reams	- - - -	1 bundle, *Bdl.*
10 reams	- - - - -	1 bale, *Bl.*

12 skins of parchment -	1 dozen, *doz.*
5 dozen - - - - - -	1 roll, *rl.*

Observe the denominations—Quire, *qr.*; ream, *Rm.*; bundle, *Bdl.*; bale, *Bl.*; dozen, *doz.*; roll, *rl.*

Two dozen *sheets* one *quire* will take,
And 20 quires one *ream* composes;
2 reams we in a *bundle* make,
10 bundles a full *bale* encloses.

By *dozens* parchment-skins are told;
12 to the dozen, as of old;
5 dozen for a *roll* are sold.

The different sizes of paper are—Foolscap, post, demi, medium, royal, super-royal, imperial, and elephant. Larger papers are described by their length and breadth in inches; thus, 20 by 32, 24 by 38, 26 by 40, 29 by 44, &c., &c.

---

2. SIZES OF BOOKS.

TABLE OF BOOKS.—This table is used by Authors, Printers, and Booksellers, in ascertaining and naming the sizes of books.

TABLE.

1 sheet of paper folded into 2 leaves is a folio, *Fol.*
1 sheet of paper folded into 4 leaves is a quarto, 4*to.*
1 sheet of paper folded into 8 leaves is an octavo, 8*vo*
1 sheet of paper folded into 12 leaves is a duodecimo, 12*mo.*
1 sheet of paper folded into 18 leaves is an octodecimo, 18*mo.*

Observe the denominations—Folio, *fol.*; quarto, 4*to.*; octavo, 8*vo.*; duodecimo, 12*mo.*; octodecimo, 18*mo.*

Whoever with a book engages,
Well knows each leaf will count 2 pages;
One *folio* sheet 2 leaves will rate,
A *quarto* 4, *octavo* 8;
A *duodecimo** a dozen clean,
An *octodecimo* eighteen;
Vicessimo quarto makes up twenty-four.
Tricessimo secundo thirty-two, no more.

---

* The duodecimo, octodecimo, vicessimo quarto, and tricessimo secundo, which are Latin numerals, are generally called 12mo., 18mo., 24mo., and 32mo. There are also 48mo., 64mo., and 72mo.

V. MISCELLANEOUS TABLES.

1. TABLE OF WEIGHTS.

A barrel of flour weighs	196 lbs.
A barrel of beef or pork	200 "
A barrel of pot ashes	200 "
A firkin of butter	56 "
A bushel of salt	56 "
A peck of salt	14 "
A gallon of train oil	7½ "
A stone of wire weighs	10½ "
A stone of sheet iron, or horseman's weight	14 "
A quintal of fish	100 "
A faggot of steel	120 "
A fother of lead	2184 "

2. TABLE OF PARTICULARS.

12 things make	1 dozen, *doz.*
12 dozen, or 144	1 gross, *gro.*
12 gross or 1728	1 great gross, *g. gro.*
20 things	1 score.
5 score	1 hundred, *C.*

3. VALUE OF SILVER AND GOLD COINS.

	$	cts.
An English shilling		24
" " crown	1	20
" " sovereign	4	84
" " pound	4	84
" " guinea	5	00
A franc of France		19
A thaler of Germany		67
A Spanish doubloon	16	00
A South American doubloon	15	60
Four shillings and two pence sterling	1	00

4. OLD ENGLISH COINS.

	s.	d.
A groat		4
A tester		6
A noble	6	8
An angel	10	0
A mark	13	4

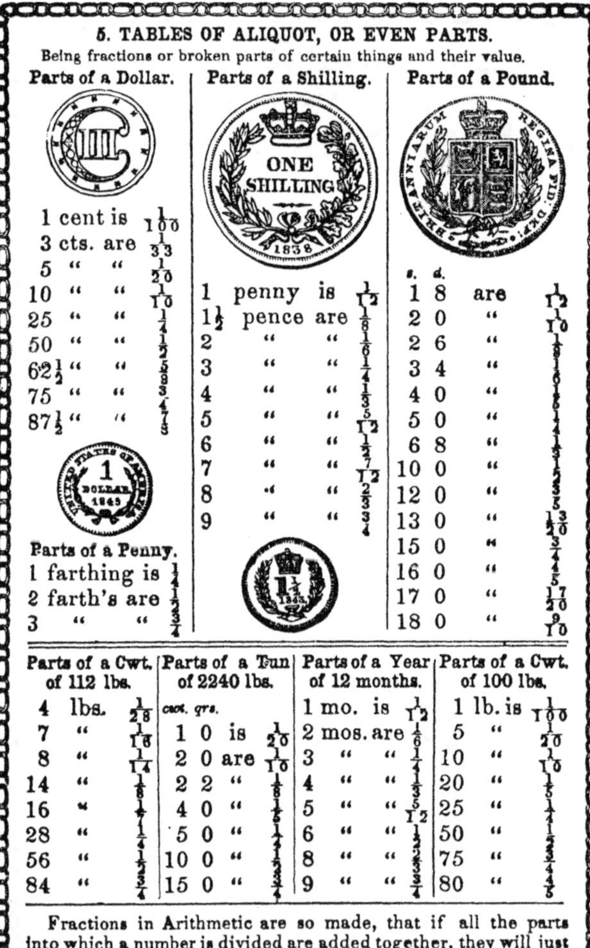

5. TABLES OF ALIQUOT, OR EVEN PARTS.

Being fractions or broken parts of certain things and their value.

Parts of a Dollar.

1 cent	is	$\frac{1}{100}$
3 cts.	are	$\frac{1}{33}$
5 "	"	$\frac{1}{20}$
10 "	"	$\frac{1}{10}$
25 "	"	$\frac{1}{4}$
50 "	"	$\frac{1}{2}$
62½ "	"	$\frac{5}{8}$
75 "	"	$\frac{3}{4}$
87½ "	"	$\frac{7}{8}$

Parts of a Shilling.

1 penny	is	$\frac{1}{12}$
1½ pence	are	$\frac{1}{8}$
2 "	"	$\frac{1}{6}$
3 "	"	$\frac{1}{4}$
4 "	"	$\frac{1}{3}$
5 "	"	$\frac{5}{12}$
6 "	"	$\frac{1}{2}$
7 "	"	$\frac{7}{12}$
8 "	"	$\frac{2}{3}$
9 "	"	$\frac{3}{4}$

Parts of a Pound.

s.	*d.*		
1	8	are	$\frac{1}{12}$
2	0	"	$\frac{1}{10}$
2	6	"	$\frac{1}{8}$
3	4	"	$\frac{1}{6}$
4	0	"	$\frac{1}{5}$
5	0	"	$\frac{1}{4}$
6	8	"	$\frac{1}{3}$
10	0	"	$\frac{1}{2}$
12	0	"	$\frac{3}{5}$
13	0	"	$\frac{13}{20}$
15	0	"	$\frac{3}{4}$
16	0	"	$\frac{4}{5}$
17	0	"	$\frac{17}{20}$
18	0	"	$\frac{9}{10}$

Parts of a Penny.

1 farthing	is	$\frac{1}{4}$
2 farth's	are	$\frac{1}{2}$
3 "	"	$\frac{3}{4}$

Parts of a Cwt. of 112 lbs.

4 lbs.	$\frac{1}{28}$
7 "	$\frac{1}{16}$
8 "	$\frac{1}{14}$
14 "	$\frac{1}{8}$
16 "	$\frac{1}{7}$
28 "	$\frac{1}{4}$
56 "	$\frac{1}{2}$
84 "	$\frac{3}{4}$

Parts of a Tun of 2240 lbs.

cwt.	*qrs.*		
1	0	is	$\frac{1}{20}$
2	0	are	$\frac{1}{10}$
2	2	"	$\frac{1}{8}$
4	0	"	$\frac{1}{5}$
5	0	"	$\frac{1}{4}$
10	0	"	$\frac{1}{2}$
15	0	"	$\frac{3}{4}$

Parts of a Year of 12 months.

1 mo.	is	$\frac{1}{12}$
2 mos.	are	$\frac{1}{6}$
3 "	"	$\frac{1}{4}$
4 "	"	$\frac{1}{3}$
5 "	"	$\frac{5}{12}$
6 "	"	$\frac{1}{2}$
8 "	"	$\frac{2}{3}$
9 "	"	$\frac{3}{4}$

Parts of a Cwt. of 100 lbs.

1 lb.	is	$\frac{1}{100}$
5 "		$\frac{1}{20}$
10 "		$\frac{1}{10}$
20 "		$\frac{1}{5}$
25 "		$\frac{1}{4}$
50 "		$\frac{1}{2}$
75 "		$\frac{3}{4}$
80 "		$\frac{4}{5}$

Fractions in Arithmetic are so made, that if all the parts into which a number is divided are added together, they will just equal that number; as when an apple is cut into parts of various shapes, you can join them together and form the apple again.

6. TABLES OF STERLING CURRENCY.

Table of Shillings and Pence.

s.		d.	d.		s.	d.
1	is	12	20	are	1	8
2	are	24	30	"	2	6
3	"	36	40	"	3	4
4	"	48	50	"	4	2
5	"	60	60	"	5	0
6	"	72	70	"	5	10
7	"	84	80	"	6	8
8	"	96	90	"	7	6
9	"	108	100	"	8	4
10	"	120	110	"	9	2
11	"	132	120	"	10	0
12	"	144	130	"	10	10

Table of Shillings and Pounds.

s.		£	s.	s.		£	s.
20	are	1	0	140	are	7	0
30	"	1	10	150	"	7	10
40	"	2	0	160	"	8	0
50	"	2	10	170	"	8	10
60	"	3	0	180	"	9	0
70	"	3	10	190	"	9	10
80	"	4	0	200	"	10	0
90	"	4	10	210	"	10	10
100	"	5	0	220	"	11	0
110	"	5	10	230	"	11	10
120	"	6	0	240	"	12	0
130	"	6	10	500	"	25	0

METRICAL SYSTEM OF WEIGHTS AND MEASURES.

WEIGHTS.

The Gram is equal to $15\frac{432}{1000}$ grains Avoirdupois.

1 Decigram is - - - - - One-tenth Gram.
1 Centigram - - - - - One-hundredth Gram.
1 Millogram - - - - - One-thousandth Gram.
1 Dekagram - - - - - Ten Grams.
1 Hectogram - - - - - One Hundred Grams.
1 Kilogram - - - - - One Thousand Grams.
1 Myriagram - - - - - Ten Thousand Grams.
1 Quintal - - - - - One Hundred Thousand Grams.
1 Millier - - - - - - One Million Grams.

LONG MEASURE.

The Metre is equal to $39\frac{37}{100}$ inches.

1 Decimetre is - - - - - - - One-tenth Metre.
1 Centimetre - - - - - - - One-hundredth Metre.
1 Millimetre - - - - - - - One-thousandth Metre.
1 Dekametre - - - - - - - Ten Metres.
1 Hectometre - - - - - - - One Hundred Metres.
1 Kilometre - - - - - - - One Thousand Metres.
1 Myriametre - - - - - - - Ten Thousand Metres.

CUBIC MEASURE.

The Litre is equal to 1 Cubic Decimetre, that is $\frac{908}{1000}$ Quart Dry Measure, or $1\frac{56}{1000}$ Quarts Liquid Measure.

1 Decilitre is . - - - - - - One-tenth Litre.
1 Centilitre - - - - - - - One-hundredth Litre.
1 Millilitre - - - - - - - - One-thousandth Litre.
1 Dekalitre - - - - - - - Ten Litres.
1 Hectolitre - - - - - - - - One Hundred Litres.
1 Kilolitre - - - - - - - One Thousand Litres.

METRICAL SYSTEM.

SQUARE MEASURE.

1 Centare is	- 1 Square Metre,	equal to		1550	Square Inches.
1 Are - - -	100	"	"	" $119\frac{6}{10}$	Square Yards.
1 Hectare - -	10000	"	"	" $2\frac{471}{1000}$	Square Acres.

TABLE OF ARITHMETICAL SIGNS.

+ ***Plus***, or more, meaning added to. This sign when placed between two numbers, shows that they are to be added together and considered as one number; thus, 24+36; that is, 24 added to 36, which is read 24 plus 36.

= ***Æqualitas***, or equality, meaning equal to. This sign when placed between two or more numbers, shows that those which precede the sign are equal to those which follow it; thus, 24+36=60; that is, 24 added to 36 are equal to 60.

— ***Minus***, or less, meaning subtracted from. This sign, when placed between two numbers, shows that one is to be taken from the other; thus, 84—42; that is, 42 is to be subtracted from 84, which is read 84 minus 42.

× ***Multiplico***, to multiply, meaning multiplied by. This sign placed between two numbers signifies that one is to be multiplied by the other; thus, 24×36; that is, 24 is to be multiplied by 36, which is read 24 multiplied by 36.

÷ ***Divido***, to part, meaning divided by. This sign placed between two numbers shows that one is to be divided by the other; thus, 60÷15; that is, 60 is to be divided by 15. When placed horizontally between two numbers, the one above and the other below, the dots are dispensed with; thus, $\frac{1}{2}$; that is, 1 divided by 2.

EXAMPLES.

Read the following: 72+47=119; 656+809=1465; 1400+700=2100; 76—38=38; 104—26=78; 290—145=145; 25 × 12=300; 99 × 9=891; 425 × 50=21250; 64÷8=8; 144÷12=12; 1728÷12=144; 1728+144—576×12÷288=54; 64÷8×16=128; 36×12=144×3; 500×10—300=47×90+470; 4—2×6=12.

——— ***Vinculum***, or a bond of union, meaning that the numbers over which it is placed, are to be considered as united or one, and to be subjected to the same operation; thus, $\overline{12+13\times19}$.

: :: : *Proportio*, or proportion. These signs are placed between numbers to show their relation to each other, so that knowing the relation of two numbers, two others may be found that have the same relation; thus, 3 : 6 :: 9 : 18, which is read as follows: As 3 is to 6 so is 9 to 18, because 3 is the half of 6 and 9 half of 18, therefore the proportion of 3 to 6 is the same as 9 to 18. This sign is read thus: As, is to, so is, to.

* *Potentia*, or power. Every number is a power, and to increase its power you must multiply it by itself the number of times answering to the power to which you wish to raise it. The figure which expresses the power to which another is to be raised is called an exponent or index, which is a small figure placed against its right hand upper part; thus, 4^3; that is, 4 is to be multiplied three times by itself, or 4×4×4=64; therefore, 64 is the third power of 4, and is the same as 4^3.

√ *Radix*, or root, meaning that the root is to be extracted. The root of a number is such a one as multiplied by itself a given number of times will produce the number whose root is wanted. √ means the square root, $^3\surd$ the cube root, $^4\surd$ the fourth root, $^5\surd$ the fifth root, &c.; therefore, √64 shows that 8 is wanted, because 8×8=64. $^3\surd 64$ shows that 4 is wanted, because 4×4×4=64. $^4\surd 16$ shows that 2 is wanted, because 2×2×2×2=16. $^5\surd 243$ shows that 3 is wanted, because 3×3×3×3×3=243.

ILLUSTRATIVE EXAMPLES.

ODD NUMBERS.

Odd numbers begin with 1, and consist of every second following figure; thus, 1, 3, 5, 7, 9, 11, 13, 15, 17, 19, are odd numbers.

Odd numbers × by odd numbers = odd numbers.
Odd numbers × even numbers = even numbers.
Even numbers × even numbers = even numbers.

Even numbers + even numbers = even numbers.
Odd numbers + odd numbers = even numbers.
Odd numbers + even numbers = odd numbers.

Many of the odd numbers above 3 that can only be ÷ 1, can be ÷ 6 by subtracting 1 or adding 1. For instance, 13 can only be ÷ 1 but 13 − 1 may be ÷ 6, so with 17, 19, 25, &c.

WHAT IS A TRILLION? a thousand billions; written thus—

1,000,000,000,000.

But if you were to count 200 a minute, it would take you 9512 years 34 days 5 hours and 20 minutes to count it, which is nearly twice as long as the world has existed

THE NUMBER NINE.

The powers of the figure 9 are more numerous and remarkable than those of any other figure.

The figures composing the product of every figure from 1 to 9 multiplied into 9, added together, make NINE; thus—

9 × 1 = 9	9 × 4 = 36. 3+6=9	9 × 7 = 63. 6+3=9
9 × 2 = 18. 1+8=9	9 × 5 = 45. 4+5=9	9 × 8 = 72. 7+2=9
9 × 3 = 27. 2+7=9	9 × 6 = 54. 5+4=9	9 × 9 = 81. 8+1=9

The above multipliers from 1 to 9 added together=45 and 4+5=NINE. Their several products added together=405, which ÷ 9=45 and 4+5=NINE. The amount of the first product (9) added to the remaining eight products (eight 9s)=81, and 8+1,=9 and 81=9×9; 81 is therefore called the *square* of NINE. The 405 mentioned above + 81=486, and this ÷9=54, and 5+4=NINE. The number of changes that may be rung on 9 bells is 362,880, which figures added together make 27, and 7+2=NINE. And 362,880÷9=40,320, and these figures added together make NINE.

THE GAME OF CHESS.

Sessa, who invented the game of chess for an East Indian king, was promised in return any reward which he should ask. Understanding, better than his patron, the power of numbers, and wishing to rebuke his rashness, he asked simply for one grain of wheat for the first square on the chess-board, two for the second, four for the third, and so doubling to the sixty-fourth. The king was astonished at the apparent smallness of the gift, but still more so, when told that the number of grains would be 18,446,744,073,709,551,615. There are in a bushel 589,824 grains of wheat, and 18,446,744,073,709,551,615 ÷589,824 gives 31,274,997,411,295 bushels; more than the whole surface of the earth could produce in many years, and more in value probably than the whole wealth of the globe The king, therefore, did not keep his promise!

GEORGE BIDDER.

This wonderful boy, whose portrait is on the title-page when very young and uneducated, could solve the most difficult arithmetical questions entirely in his own mind without the use of pencil or slate, and more quickly than any one could in the common way.

www.ingramcontent.com/pod-product-compliance
Lightning Source LLC
LaVergne TN
LVHW011123110826
845150LV00008B/2233

* 9 7 8 1 4 1 8 1 9 5 1 0 6 *